Science in Faith and Hope
an interaction

George Ellis

Q Quaker Books, 2004

The talk on which the text of this pamphlet is based was organised under the auspices of Quaker Quest, a meeting held each Monday evening at Friends House for people to find out more about Quakerism. A recording was published by Quaker Quest as a CD and cassette tape, available from the Quaker Bookshop, 173 Euston Road, London NW1 2BJ

Editor: Deborah Padfield

Published December 2004 by Quaker Books

© George Ellis 2004   ISBN 0 85245 371 X

The moral rights of the author are asserted in accordance with the Copyright, Designs and Patents Act 1988. All rights reserved. No part of this book may be reproduced or utilised, in any form or by any means, electronic or mechanical, without permission in writing from the publisher. Reviewers may quote brief passages. Enquiries should be addressed to the Publications Manager, Quaker Books, Friends House, 173 Euston Road, London NW1 2BJ

Contents

Introduction *by Peter Eccles* 5

The Conflict 7

The Universe 8

Evolution 10

The Nature of Existence 11
    The Anthropic Question 11
    Chance 13
    Design 13
    The Multiverse 13

The Nature of Humanity 15
    Top-Down Causation 16
    Effectiveness of Ethics 17
    Physics and Purpose 18

Limits to Science 19
    Science and Ethics 20
    Aesthetics, Metaphysics, Meaning 22

Emotions and Values, Faith and Hope 22
    Balancing Reason, Emotion, Ethics 23

Moral Realism and the Nature of Ethics 24
    Kenosis 24
    Universality and Effectiveness of Kenosis 28

Fundamentalism and Humanity 30
    Scientism 30
    Reductionism and Humanity 34
    Neo-Behaviourists and the Mind 35

Existence and Evidence 38
    Image and Reality 38

The Wider Nature of Kenosis 39
    Discernment 40
    The Inner Voice 40

The Issue of Evil 41

The Whole Picture 42
    Awareness of Transcendence 42
    Embracing the Whole 43

Bibliography 43

Introduction

The text of this pamphlet originated in a talk given by George Ellis at Friends House in London on 4th May 2004, at a meeting organised by Quaker Quest to enable interested people to learn more about Quakerism. The talk has been extensively edited and extended to incorporate points made during questions at the end. Peter Eccles introduced George Ellis as follows:

George Ellis has been awarded the Templeton Prize for progress towards research in or discoveries about spiritual realities, including research in love, creativity, purpose, infinity, intelligence, thanksgiving and prayer. This is a valuable annual prize awarded by the Templeton Foundation. It was first awarded in 1973 to Mother Teresa, and George Ellis is the 34th prizewinner.

George Ellis graduated from the University of Cape Town in 1960. Having come to Cambridge in England for postgraduate study, he embarked there on his academic career. He is now working in the Department of Mathematics at the University of Cape Town where he's the Distinguished Professor of Complex Systems, also holding a Visiting Chair in Astronomy at Queen Mary College in London. His scientific work has mainly been in the application of general relativity theory to cosmology. Cosmology is about the origin and evolution of the universe, and in 1973 with Stephen Hawking he published a famous book, *The Large Scale Structure of Space Time*.

He returned to South Africa in 1973, to Cape Town, but then he got involved in other things there. He was critical of the Nationalist Government and he was particularly interested in homelessness and housing policy. He became a Quaker in 1974. George Ellis's involvement from about fifteen years ago in issues of science and religion has in certain ways built on his experience in South Africa, and has led to this very prestigious award.

Science in Faith and Hope: an interaction

The Conflict

You will all be aware that there has for some time been a view that science and religion are in conflict with each other. This view has been propagated strongly by some people, and in a sense each topic has at some time in the past claimed more than it should have, thereby indeed creating such a conflict. However, that is now changing.

At the beginning, religion was claiming to provide "scientific" texts on the origin of life, the physical origin of the universe, and the like. As science has progressed, religion has had to pull back from making historical claims about the evolution of life, the history of the universe, and so on. Science has since then been claiming more and more territory, providing scientifically based explanations of ever wider aspects of the world and the universe; now some scientists are claiming that science explains everything—indeed that it will provide all the understanding one needs to live an adequate life. They are, of course, wrong. They are also going to have to pull back.

There's nothing new in the idea that religion has its place in dealing with meaning, ethics, and metaphysical issues, while science has its disparate place in dealing with mechanisms—with how things work. If one looks at it in this way, there is not much conflict between them, as they each deal with separate domains.

Some people have even tried to claim that because they deal with such different domains, there simply is no room for any conflict—there are non-overlapping magisteria of interest. That is not quite right. There are some places where there are at least potential conflicts.

The Universe

In the past, one conflict concerned the origins of the universe. There is no reason nowadays to question that the universe expanded from an early hot Big Bang epoch with only elementary particles present and with matter and radiation in equilibrium. This epoch ran from, let's say, a temperature of 10^{12} degrees—1 followed by 12 zeros—down to the present day. During this time, there were processes like nucleosynthesis (creation of the lightest elements), the decoupling of matter and radiation as they cooled, the formation of early stars and galaxies, followed by the occurrence of supernova explosions which spread heavy elements through space, with the remnants forming the seeds for second generation stars and planets. All this is pretty much understood. At earlier times than 10^{12} degrees, there probably was a very rapid era of accelerating expansion—an epoch called *Inflation*—that made the universe very large and smooth. But the issue of what happened before even then remains unresolved at present. Present day science by and large supports the view that there was indeed a start to the universe, with various proposals as to how this might have happened, but with some theories of the quantum origin of the universe suggesting otherwise. The crucial issue is, would it be bad for religion if there was no beginning to the universe?

It used to be believed by some that if you could prove the universe had a beginning, this would vindicate biblical claims and so would be good for religion. On the other hand, if you could prove that the universe did not have a beginning, as with Fred Hoyle's theory of the steady-state universe, this would undermine a religious worldview.

However, even in the time of St Augustine it was known that this was a faulty analysis, for the crucial existential issue is not dependent on whether the universe had a beginning in time. The fundamental question is why the physical universe has the form it does—why the laws of physics have the specific nature they have, and why the expanding universe has the specific characteristics it has. That issue remains a fundamental metaphysical question, whether the universe had a start or not. Why does the universe have this particular form, when it could in principle have had so many other forms?

Most scientifically-oriented theologians working in this area feel quite comfortable with belief in a creator God as in John's Gospel, Chapter One, whether there was a start to the universe or not. God could have created the universe in many different ways, with or without a starting event, and the way she chose to create it is a matter of scientific interest but has no real theological substance. That is simply a question of what mode God chose to use to bring the universe into existence and maintain it in being; the creative power of God remains, whatever the physical vehicle chosen.

Actually present day science does indeed seem to support the idea that there was a beginning to the universe. However, the question of the origin of the universe is very tricky from a scientific viewpoint. The "place" where the origin happened is not a regular part of spacetime, it is the boundary of space and time. That means it is not subject to the same physical laws. The whole concept of the Big Bang is extraordinary because it is not just that matter comes into existence: space and time also come into existence. The title of my book *Before the Beginning* is an intentional conundrum, because when you try to work out what happened "before the beginning", you're already talking nonsense. There wasn't any before.

It is very hard to cope with that in our language, which is based on our own experience in the existing universe. People are developing all kinds of theories about this. These are largely untestable theories; they are often sophisticated, but not all are self-consistent. We do not know what happened then, and we have no compelling theories. Known science breaks down at that point, and perhaps even more importantly our ability to test science breaks down. So anything anybody says when proposing theories of creation of the universe is hypothetical. It may or may not have been so.

Evolution

The more controversial question concerns the origins of life, the mechanism of the evolution of animals and humans. A huge amount has been written about what is claimed to be the upsetting of the religious view by Darwinian evolutionary theory. And indeed the old religious view—crudely speaking, God sitting at the drawing board devising giraffes and zebras and lions and so on—has gone by the board. Instead, with the modern view of evolution, what you have is an understanding of the incredible self-structuring propensity of nature, which enables the Darwinian evolutionary process to shape animals to function well in their environment.

This works through a process of random variation in the detailed nature of animals, followed by selection of those animals best suited to survive in their environment, simply because more of their progeny survive than do those of the less fit. The best adapted animals pass on their genes to their children, so that the species as a whole becomes better and better adapted to their environment. This leads to greater and greater complexity, including consciousness, because the ability to think enhances survival capacity.

Now this self-structuring propensity is based in the laws of physics. In particular, it is based in the way electro-magnetism and quantum theory work. These underlie chemistry, chemistry underlies biochemistry, biochemistry underlies the way that cells function and life comes into being. From the modern viewpoint, if God chose to create humans by the process of designing laws of physics which then makes the coming into being of life inevitable, well, that's a wonderful way of doing it. There's nothing wrong with that at all from a theological perspective.

So despite people in the rearguard still fighting out-of-date battles about this issue, in my view there is no real conflict here. If a creator shapes laws of physics of such a nature that they ensure that life will come into being, that is an amazing way of getting creation going. There is no serious theological problem, except that you may start worrying about the suffering involved in it, but that is part of the bigger problem of suffering which arises independent of the nature of evolutionary theory. I will briefly return to that later.

The Nature of Existence

But there are still some major questions where potential or real conflict can arise. The first is the issue of existence. Why is there a universe? Why are there any laws of physics? Why are the laws of physics the way they are?

The Anthropic Question

One particular important issue arising here is the *anthropic question*. The "anthropic principle" comes in various forms, each attempting to relate the existence of life in the universe to the nature of the universe: see *The Anthropic Cosmological Principle* by John Barrow and Frank Tipler. The way life evolves depends on the universe. We can consider universes with all sorts of properties: bigger or smaller; hotter or colder; expanding faster or slower; with different laws of physics, different kinds of particles, different masses for particles; maybe with different laws of physics altogether. As a cosmologist, one imagines all these different universes and considers what they would be like. In most of them there would be no life at all.

What is clear is that life as we know it would not be possible if there were very small changes to either physics or the expanding universe that we see around us. There are many aspects of physics which, if they were different, would prevent any life at all existing. There are all sorts of subtleties if the whole thing is to work, allowing complexity to emerge: for instance, the difference in mass between the proton and the neutron has to lie in a very narrow range, and the ratio of the electro-magnetic to the gravitational force has to be very finely tuned. If you tinker with physics, you may not get any element heavier than hydrogen; or maybe if the initial conditions of the universe are wrong it doesn't last long enough, or it's always too hot, or it expands so rapidly that no stars form at all. So there are all sorts of things that can go wrong if you are the creator trying to create a universe in which life exists. We are now realising that the universe is a very extraordinary place, in the sense that *it is fine tuned so that life will exist*. Because of its specific nature, our existence is more or less inevitable: since its start, it was always waiting for us to come into being (as well as other intelligent beings elsewhere in the universe).

A lot of books have been written on this, for instance *Just Six Numbers* and *Our Cosmic Habitat* by my friend and colleague Professor Martin Rees, the Astronomer Royal. In particular, it has recently been established that the universe is not at present slowing down as we would have expected, but rather is accelerating. It is expanding faster and faster due to a cosmic force known as the "cosmological constant" or "quintessence": various names are used for it. We do not know why this force is there, but we do know that if it were substantially bigger than it is, there would be no galaxies at all, no planets, no life at all. The question is why it has the value it has, when fundamental physics suggests it should have been hugely larger than it in fact is, with no structure at all forming in the universe and hence no life.

So why does the universe allow life to exist? Some scientists do not see this as a valid issue, but in my view it is a very serious question. There are three main ways of trying to explain it.

Chance

The first is to invoke pure chance. Just by chance everything worked out right. This is not saying that what happened is likely; rather, it is just what happened, and there is nothing more to say. That is a logically tenable position, if you like to live with extremely thin philosophies. But it has no explanatory power; it doesn't get you anywhere. So it is not an argument that is popular in many scientific or philosophical circles.

Design

The second option is the good old designer argument: that the way the universe functions reveals intention, the work of some kind of transcendent power or force. Life exists because this fine-tuning intentionally took place. In simple terms, God designed the universe and the laws of physics operational in it in such a way that it was inevitable that life would come in to being. Physics has this extraordinary ability to underlie the spontaneous development of complexity because that was intended to be the case. This is the theistic view.

The Multiverse

The third option is the idea of a multiverse, supported amongst others by Professor Martin Rees, who has studied and written about the problem of fine-tuning in the books I mentioned. He and others propose that this is not the only universe, but that there are millions and millions or even an infinite number of other universes, all with differing properties. Or perhaps there is just one huge universe with many, many different expanding regions like the expanding universe region we can see around us, but each with different physical constants, different rates of expansion, and so on. This enables physicists to start doing what they like, talking about statistics of universes.

If you play the game right, you can say that in this context life is highly probable. In most of these universes life will not occur because conditions will be wrong. But in a few of them it will just happen to work out alright. So although there is an incredibly small probability of a universe existing that will allow life, if there exist enough universes with all sorts of properties occurring, it becomes essentially inevitable that somewhere the right mix of circumstances will occur. There are all these zillions of universes, so in some of them surely life will come into being.

The problem with this explanation is that none of these other universes or expanding universe domains can be observed. They are beyond the part of the universe that we can see, so whatever is said about them can never be proven wrong. In many cases, there is no causal connection with them whatever, so there is not even the faintest possibility of checking their existence or their properties. That makes this a metaphysical rather than a scientific proposal.

The distinguishing feature of science is that you can test its proposals, and there is no way of testing this hypothesis. People are trying to show that you could test it in some way, but in my view they are failing. It seems to me that belief in the existence of a multiverse requires the same amount of faith as belief in a God who creates one universe. Either option is possible. We cannot prove without doubt that either is correct. Martin Gardner has written about this interestingly in a book called *Are Universes Thicker Than Blackberries?*—concluding that there is no evidence whatsoever that there is more than one universe.

A final point is that if there indeed were a multiverse, that would not necessarily exclude a creator God: for she could have decided to create many universes instead of one. These are not in fact exclusive options. And indeed the multiverse proposal is not after all a final explanation; it just pushes the final question back one stage further. The issue becomes, why this multiverse rather than that? Why a multiverse that allows life, rather than one that does not? The crucial question recurs.

The Nature of Humanity

The second major current issue is where the real crunch comes. It is the question of what is the essential nature of humanity in the light of modern biology, in particular molecular biology and neuroscience. This is where there is real potential for conflict between science and religion, which is going to go on for a long time.

On a scientific basis, we have been obtaining greater and greater understandings of how life works. In biology, we are learning more about the way that molecular processes underlie the functioning of each of us, particularly through DNA as the store of our genetic heritage and through neurons as the basis for our minds. We have been gaining a remarkable mechanistic understanding of the way that life works, which is extraordinarily successful. The problem arises when the claim is made that this is all there is: nothing else is relevant or has any reality.

Here we come up against the fundamentalist views of reductionists who produce incredibly thin views of humanity. It is an extraordinary phenomenon: people from sociology, psychology, evolutionary theory, molecular biology, neuroscience, philosophy, and so on, making claims that humans are far less than they actually are. They do so with great authority, and if you disagree with them on humanistic or religious grounds you are greeted with derision. This is a really important area.

Underlying humanity is the basic physical hierarchical structuring: quarks making up protons and neutrons, which with electrons make up atoms; atoms together make molecules; enough molecules together make bio-molecules; if you string these together you eventually get cells; cells make tissues, tissues make systems, systems make the organism, and organisms make communities. This is the physical structure underlying our existence and functioning.

Top-Down Causation

A common view is that the only causal effect present in this hierarchy is bottom-up causation, the attraction between tiny electrons and protons at the bottom causing everything all the way up. In a certain sense that is obviously true. You are able to think because electrons are attracting protons in your neurons. Yet the new, very important realisation is that as well as the bottom-up action, there is top-down action. In this hierarchy of structure, the top levels are equally able to influence what happens at the lower levels. It is this simple fact that somehow quite a lot of people seem to miss. Top-down action is what enables the higher levels of the hierarchy to be causally effective, and it occurs around us all the time. This happens for example in the physics of the very early universe, in the way that genes are influenced by the environment through the process of evolution, in the way that genes are read on the basis of positional information in the developing body, and it happens in the way that the mind influences what happens in our bodies.[1]

A key point here is the issue of human volition: the fact that when I move my arm, it moves because I have "told it" to do so. In other words, my brain is able to co-ordinate the action of millions of electrons and protons in such a way that it makes the arm move in the way I want it to move. Everything in this room that is created by humans is created through human volition, and this demonstrates that our minds are causally effective in the world around us. The goals that we have shape what we do, and hence change the situation in the world around us as we act on it, for example when building a house or driving a car or making a cup of tea. It is this causal efficacy that enables us to carry out actions embodying our intentions. This in turn is crucial to our existence as ethically responsible beings.

It is important to understand that information is causally

---

[1] For a discussion of these effects, see: www.mth.uct.ac.za/~ellis/emerge.doc

effective, even though it is not a physical quantity but rather has an abstract nature. It is because of this effectiveness that it costs money to acquire information—it has an economic value. Not only that, but social constructions, too, are causally effective. The classic example of this is the laws of chess. Imagine someone coming from Mars and watching chess pieces moving. It is a very puzzling situation. Some pieces can only move in one way and other pieces can only move another way, so you imagine the Martian turning the board upside down or looking inside the rook, searching for a mechanism that causes these differences.

But it is an abstraction, a social agreement, that is making the chess piece move only in these ways. Such an agreement, reached by social convention over many hundreds of years, is not the same as any individual's brain state, though some people will try to tell you that it is. It is an abstraction that exists independent of any single mind, and that can be represented in many different ways. It is causally effective through the actions of individual minds, but none of them by themselves created that abstraction or embody it in its entirety. It will survive even when they die.

Many other social constructions, including language, mathematics, and science, are equally causally effective. This already is enough to undermine any simplistic materialistic views of the world, because these causal abstractions do not have a place in the simple materialist view of how things function. Indeed materialism itself is a theory, but a theory is not a physical thing. Its very existence denies its own fundamental premise.

Effectiveness of Ethics

In a similar way, ethics too is causally effective. It is the highest level of the goals we have, because ethics is the choice of which other goals are acceptable. When you have chosen your ethical values, your value system, this governs which goals are inside your acceptable boundary and which are outside. So this choice is, again, causally effective. As a simple example, if your country believes that the death penalty is okay, this will result in the physical realisation of that belief in the physical existence of an electric chair or some equivalent in your jails. If your country does not believe in the death penalty, they will not be there.

The embarrassing thing is that this causal effectiveness of ethics is obvious to the man and woman in the street. I keep emphasizing this feature because it lies outside what materialist, reductionist physicists and chemists have in their causal schemes. But as soon as I say it, it is obviously true. It is crucial to a religious view of life. And an extraordinary thought is that ethics can be causally effective only because of the detailed nature of physics, because physics is controlling the way that electrons and molecules flow in our neurons and enable our thoughts.

So one of the really interesting questions is, which aspect of physics is it which allows this to happen? And how different could physics be and still allow this? This is the study which leads to the anthropic conundrum that I have already mentioned.

Physics and Purpose

A further important point about physics in relation to the science and religion debate is that, despite what any physicist may explicitly tell you or implicitly imply, physics as it currently stands is causally incomplete. It is not able to describe all the forces in action shaping what happens in the world around us. I like to illustrate this with a pair of spectacles.

Physics cannot explain the curve of the glass in the spectacles, because they have been shaped on purpose to fit my individual eyes. Now, the hierarchy of variables considered by physics contains quantities like momentum and energy, pressures and densities, and so on. But the vocabulary of physics has no variable corresponding to intention. Because of this, physics cannot explain why these spectacles have their particular curvature. This is not a statement of vitalist belief. It is a simple statement of fact, which remains true irrespective of what your philosophy of mind may be.

This means that physics today provides a causally incomplete theory of the world around us. It cannot describe all the causes acting to shape what happens in the real world. However, you will not find this said in any of the physics textbooks in bookshops. For physics to be causally complete, at a minimum it would have to introduce some variables corresponding to intentions and rational plans. That would in turn involve introducing equations attempting to show how intentions form, in order to explain the curvature of the spectacles, the existence of motor cars, the structure of computers, and indeed the existence and nature of the building in which we are presently situated. None of these variables or equations exist in present day physics, which therefore cannot comprehend the corresponding effects.

Limits to Science

What I am talking about here is one of the limits of present day science. This is a boundary that might conceivably change one day (although I very much doubt it). But there are also many limits to what science will ever be able to do, that will never change—they are boundaries to its competence, because of its nature and its methods of investigation.

This is a really important issue. There are many areas of concern for humans, of which only a subset are within the ambit of science. Outside of this ambit are crucially important areas: in particular, ethics, aesthetics, metaphysics, and meaning. They are outside the competence of science because there is no scientific experiment which can determine any of them. Science can help illuminate some of their aspects, but is fundamentally unable to touch their core.

Now, when I mention these limits of science, some people immediately say, "Ah, this is the old 'God of the gaps' argument," that is, it is an attempt to provide a religious explanation for something that science will eventually explain. It is nothing of the sort. These are about absolute boundaries to what science can ever do, because of the very nature of science.

Science and Ethics

There is a great deal of confusion about this, particularly in the case of ethics. This is outside the competence of science because there is no experiment which says that an act is good or bad. There are no units of good and bad, no measurements of so many milli-Hitlers. Ethics is simply an area that science cannot handle. It is true that science depends on and supports some basic virtues such as respecting the data, telling the truth, and so on, as pointed out by Jacques Monod. This does not begin to touch real ethical issues to do with the relative importance of ends and means, how to deal with conflicting interests, how to balance outcomes against principles, and so on. They simply do not help as regards real world ethical dilemmas.

However sociobiology and evolutionary psychology produce arguments which claim to give complete explanations as to where our ethical views come from. There are many problems with those attempts, the first being they do not explain ethics, they explain it away. If the true origin of our ethical beliefs lay in evolutionary biology, ethics would be completely undermined, because once you understood this you would no longer necessarily believe that you had to follow its precepts. You could choose to buck the evolutionary imperative.

The second is that this is a typical fundamentalist argument which looks at some of the causes in operation and ignores others; it simply leaves out of consideration two other important parts of the equation, namely social effects embodied in culture (which some sociologists and anthropologists with equal fundamentalist vehemence claim are all that matter), and personal choice.

The third is that if you did follow those precepts, you would rapidly end up in very dangerous territory, namely the domain of social Darwinism. That has been one of the most evil movements in the history of humanity, causing far more deaths than any other ideology has done. And the fact that I am able to say it is evil shows that there are standards of ethics outside of those provided by evolutionary biology. Yes I know there is a substantial lit-erature on the evolutionary rise of altruism, but as a historical fact the influence of evolutionary theory on ethics in practice has been to provide theoretical support for eugenics and social Darwinism, not for any movement of caring for others. The historical record is quite clear on this: see Richard Weikart in *From Darwin to Hitler*. He demonstrates that many leading Darwinian biologists and social thinkers in Germany believed that Darwinism overturned traditional Judeo-Christian and Enlightenment ethics, especially the view that human life is sacred. Many of these thinkers supported moral relativism, yet simultaneously exalted evolutionary "fitness" (especially intelligence and health) to be the highest arbiter of morality. Darwinism played a key role in the rise not only of eugenics, but also euthanasia, infanticide, abortion, and racial extermination. This was especially important in Germany, since Hitler built his view of ethics on Darwinian principles, not on nihilism.

Challenging evolutionary biologists who still maintain that their science can provide a basis for ethics, despite these arguments, is very simple. If a scientist says, "Look, science can handle ethics," say to them, "Tell me, what does science say should be done about Iraq today? And tell me what science says ethically about Israel and Palestine?" You will get a deafening silence, because the simple fact is that science cannot handle ethical questions. Ethical values, crucial for our individual and social lives, have to come from a value-based philosophical stance or a meaning-providing religious position. They cannot be justified by rationality alone, much less by science.

Aesthetics, Metaphysics, Meaning

Similarly, aesthetics—the criteria of beauty—is also outside the boundaries of science. No scientific experiment can determine that something is beautiful or ugly, for these are not scientific concepts. The same is true for metaphysics and meaning. Thus there are major areas of life, incredibly important to humanity, which cannot be encompassed in science. They are the proper domain of philosophy, of religion, of art, and so on, but not of science.

Why are there these boundaries? Because experimental science deals with the generic, the universal, in very restricted circumstances. It works in circumstances so tightly prescribed that effects are repeatable. Most things which are of real value in human life are not repeatable. They are individual events which may have crucial meaning for individuals and for humanity in the course of history; but each occurs only once. So repeatable science does not encompass either all that is important or all that can reasonably be called knowledge.

Emotions and Values, Faith and Hope

Furthermore, we have the tension between rationality, emotion, faith, and hope. Some science-based worldviews claim in essence that reason is all that is needed for life, while emotion, faith, and hope simply get in the way of rationally desirable decisions. But this is a false view. It is not possible to reason things out and make decisions purely on a rational basis. Firstly, we need values to guide our rational decisions; but these cannot be arrived at rationally. Secondly, in order to live our lives we need faith and hope, because we always have inadequate information. It is a part of daily life that when we make important decisions like whom to marry, whether to take a new job, whether to move to a new city, they are always to a considerable degree guided by emotion, and in the end have to be concluded on the basis of partial information.

Thus a lot of choices are based on faith and hope, faith about how things will be, hope that it will work out all right. This is true even

Emotions and Values, Faith and Hope

in science. My scientific colleagues set up research groups to look at string theory or particle physics and so on. They do so in the belief that they will be able to obtain useful advances when their grant applications have been funded. They do not know that they will make those steps forward. It is a belief; it is a hope. So embedded in the very foundations even of science there is a human structure of faith and hope. Furthermore they actually carry out these enterprises because of the associated emotions and values that guide their actions. For example, the desire to understand is an emotion that underlies much of science.

Balancing Reason, Emotion, Ethics

It is crucial to understand that our minds act, as it were, as an arbiter between three tendencies guiding our actions: first, what rationality suggests is the best course of action—the cold calculus of more and less, the economically most beneficial choice; second, what emotion sways us to do—the way that feels best, what we would like to do; and third, what our values tell us we ought to do —the ethically best option, the right thing to do.

These are each distinct from each other, and in competition to gain the upper hand. Sometimes they may agree as to the best course of action, but often they will not. It is our personal responsibility to choose between them, making the best choice we can between these conflicting calls, with our best wisdom and integrity, and on the basis of the limited data available.

This shows where value choices come in and help guide our actions. Rationality can help decide which course of action will be most likely to promote specific ethical goals when we have made these value choices, but the choices themselves, the ethical system, must come from outside the pure rationality of rigorous proof, and certainly from outside science. As emphasized before, science cannot provide the basis of ethics. A deep religious worldview is crucial here. It is essential to our well-being and proper fulfilment, because ethics and meaning are deeply intertwined. Humans have a great yearning for meaning, and ethics embodies those meanings and guides our actions in accordance with them.

I am presenting here a very Quaker view of things: that one needs to bring ethics into the science and religion dialogue. In fact, the whole discussion is incomplete without ethics for many reasons. Thus it is important to have a sound view on the nature of ethics and morality.

Moral Realism and the Nature of Ethics

I take the position of moral realism, which argues that we do not invent ethics, but discover it. A whole sociological school suggests that we invent ethics—it is created by society. That route ends up in total relativism, where it is impossible to say that any act is evil, for it is a simple fact that different social groups have different ethical beliefs. All you can say from this standpoint is that some people have different socially determined values than others; nothing can be labelled as bad or evil. It is simply that Hitler and Churchill were told different things by their mothers. Neither did anything wrong, for "wrong" has no universal meaning.

If you however believe truly that some acts are good and some are bad, you have to recognise the ability to distinguish good and bad—and that is a statement of ethical realism. This means that we don't invent ethics, rather we discover it, in much the same way as we discover mathematics, which has a universal character that is the same everywhere in the universe. The proposal that Nancey Murphy and I make is that ethics also has a universal character that is invariant across time and space.

Kenosis

I am talking here about deep ethics, which is different from the shallow ethics on which everybody agrees and which sociobiology can explain. What is deep ethics? It is kenosis: self-emptying, or giving up, or self-sacrifice, which is deeply embedded in all the religious traditions of the world. It is the core of Christianity: the

Moral Realism and the Nature of Ethics

suffering on the cross is a kenotic, self-sacrificial giving up on behalf of humanity. I believe it is deeply embedded in the Quaker view of how to handle war and peace. In *Christian Faith and Practice* there is a passage (§606) about this, which goes as follows:

> The Quaker testimony concerning war does not set up as its standard of value the attainment of individual or national safety, neither is it based primarily on the iniquity of taking human life, profoundly important as that aspect of the question is. It is based ultimately on the conception of "that of God in every man" to which the Christian in the presence of evil is called on to make appeal, following out a line of thought and conduct which, involving suffering as it may do, is, in the long run, the most likely to reach to the inward witness and so change the evil mind into the right mind. This result is not achieved by war. (A. Neave Brayshaw, 1921)

Changing the hardened heart is not achieved by military force or by buying people, it is not achieved by intellectually persuading people; it is achieved by touching them as humans through treating them as valuable. It is achieved particularly by sacrifice on behalf of others, as exemplified in the life and work of Martin Luther King, Mahatma Gandhi, Desmond Tutu.

The attitude of deep ethics is not that you are *always* self-sacrificing on behalf of others, it is that you are prepared to do so if and when it will make a strategic difference. That is a different thing. There are times when it is the only thing which will make a real difference. Kenosis is when we are ready to sacrifice ourselves at the right time and place if it will be transformatory, particularly when it can change an enemy into a friend. For changing enemies into friends is the basis of true security.

This is the subject of the book I wrote with Nancey Murphy, called *On the Moral Nature of the Universe*. It has on its cover a picture of Dresden in 1945 after the firebombing by the allies. Many of you will know that picture: the burnt-out ruins of the town below, an angel in the foreground with pitying hands stretched out over Dresden. We used that picture to illustrate the ultimate alternative to a kenotic view of life. If you see the other only as an implacable enemy, it is in those deathly ruins of Dresden that you eventually end.

Many have suggested that kenosis is a nice idea, but impractical in reality. In response to that, I will show you a most remarkable document I received recently. When my views were broadcast following the award of the Templeton prize, a man called David Christie got in touch with me:

> In 1967 I was a young officer in a Scottish Battalion engaged in peacekeeping duties in Aden, in what is now Yemen. The situation was similar to Iraq with people being killed every day. As always those who suffered the most were the innocent local people. Not only were we tough but we had the firepower to pretty well destroy the whole town had we wished, but we had a commanding officer who understood how to make peace and he led us to do something very unusual, not to react when we were attacked. Only if we were one hundred per cent certain that a particular person had thrown a grenade or fired a shot at us were we allowed to fire. During our tour of duty we had 102 grenades thrown at us and in response the entire battalion fired the grand total of two shots, killing one grenade thrower. The cost to us was over 100 of our own men wounded, and, surely by the grace of God, only one killed.
>
> When they threw rocks at us we stood fast, when they threw grenades we hit the deck and after the explosion we got to our feet and stood fast. We did not react in anger or indiscriminately. This was not the anticipated reaction. Slowly, very slowly the local people began to trust us and made it clear to the "local terrorists" that they were not welcome in their area.
>
> At one stage neighbouring battalions were having a torrid time with attacks. We were playing soccer with the locals. We

had in fact brought peace to our area at the cost of our own blood. How had this been achieved? Principally because we were led by a man who every soldier in the battalion knew would die for him if required. Each soldier in turn came to be prepared to sacrifice himself for such a man.

Many people may sneer that we were merely obeying orders but this was not the case. Our commanding officer was more highly regarded by his soldiers than the general. One might almost say loved. So gradually the heart of the peacemaker began to grow in each man, in a determination to succeed whatever the cost. Probably most of the soldiers like myself only realised years afterwards what had been achieved.

I met with David Christie and talked with him. He is a soldier through and through. He is prepared to kill but believes in peace. As a result I am putting him in touch with Nosizwe Madlala-Routledge who is a member of our Cape Western Meeting and was until recently the Deputy Minister of Defence of South Africa. I am putting David Christie in touch with her in the hope that he will be used to help train the South African Defence Force.

Peace is not about flattening yourself or being flattened. It is a question of the nature of the interaction. I must not claim to be more of an expert in this than I am. But as I see it, the key point is that if we eventually have to use coercive methods to prevent slaughter of the innocent, all the time we should be offering the other person a way out, offering them their full humanity; we must not be saying "you are irredeemable". I think that is the key. All the time it must be clear to that person: "I'm going to do everything I have to do to stop you but I'm not treating you as sub-human." There must be that chance, that opportunity.

There is a whole range of positions, from "getting to yes" kind of actions, involving negotiation, conflict resolution, and empathising, through the full nature and cost of forgiveness, to the deep ethics of kenotic self-sacrifice. Forgiveness is a huge step on the way. It is not the whole way but it is part of it. The key is the ability truly to see others as fully human instead of seeing them through the enemy image which allows you to treat them as sub-human. At the deep end it involves sacrifice on behalf of our enemies even unto death, an almost impossible goal to achieve. Nevertheless this self-sacrificial or kenotic ethics is the true nature of deep ethics, which we discover rather than invent. It appears in the deep spiritual traditions of all the religious faiths.

Universality and Effectiveness of Kenosis

When I spoke about this in California some years ago, a gentleman came up to me in great excitement after the talk and said, "That was a terrific talk, you spoke like a true Muslim." I was amazed. He was the director of the Muslim Studies Centre here in London. In New York, I heard a talk given by the Chief Rabbi of Great Britain in which exactly the same spirit was expressed, and I told him, "You talk like a Quaker." He said, "I choose to take that as a compliment." The same understanding is deeply embedded for example in the Hindu tradition in which Mahatma Gandhi grew up.

Thus I believe that each of the major world religions has a spiritual tradition that believes seriously and deeply in a kenotic ethics. The real division is not in this ethical sphere. The division lies between the fundamentalists and the non-fundamentalists, irrespective of whether they are within the same religion/faith group or not. The non-fundamentalists can get on with each other irrespective of their faith positions, and, I suggest, agree on the deep nature of ethics, whichever faith they come from.

The South African struggle has influenced me a lot in understanding this. Firstly, I think the strength of my belief in the reality of moral issues comes from confronting evil in that context, and being convinced that it was indeed evil. There was no relativist way in which you could talk it away. Secondly, the opposition groups to which I belonged in South Africa, and many other groups, were extraordinarily principled. They embodied a lot of the self-

Moral Realism and the Nature of Ethics

sacrificial ethic in the way they worked.

The temptation in that South African situation, facing the evil of the Nationalist government, was to demonise the government. I found it very difficult in many cases to resist that temptation. It was hard to see the light of God in the Minister of Justice who was responsible for deaths in detention and so on. It is incredibly easy to demonise others. But it is a slippery slope: once started, you soon begin to be prepared to treat them in ways which should not be countenanced. A lot of the opposition was very careful about this, particularly of course Desmond Tutu and Nelson Mandela.

Much of this was embodied in the Truth and Reconciliation Commission process. The bargain was that forgiveness would be given even when anger was highly justified, for which terrible reparations could have been demanded. The political peace which ensued was made possible by the forgiveness engaged in by the Black populace following the leadings of Desmond Tutu, Nelson Mandela, and others. Steve Biko was a person of the same calibre.

There were cases where the TRC process worked in an amazing way. In other cases, it totally failed. Nevertheless, it was a political process of forgiveness instead of revenge. Without that we would have been engaged in a bloody civil war.

In our interaction with the world around us, everything depends on our understanding of context. If the context changes, the meaning of everything changes. Forgiveness is an interesting example. You can never change *the facts* about what has happened in the past, but you can change *their meaning* by changing your interpretation, altering the context within which you understand it. This is how kenosis can be effective, despite its paradoxical nature ("He who would save his life shall lose it, he who loses his life shall save it..."—Mark 8:35). The image of a hateful enemy is changed into the image of a suffering human being; then things are totally different. Context is crucial in this. This is what fundamentalism ignores.

Fundamentalism and Humanity

I would like to say a bit more about fundamentalism, in relation to the question of what makes us human. More and more, I am realising that this is what I am fighting across the board. What do I mean by fundamentalism? I mean *the proclamation of a partial truth as the whole truth*. This is the essential nature of the problem.

The power of fundamentalism derives from the fact that the partial truth proclaimed is indeed a truth. The danger is that it denies that any other part of the truth is also true. There are of course religious fundamentalists in each of the main religious movements, including fundamentalist atheists. But there are also fundamentalists in each area of human thought, including art and literature, human and social sciences. And in particular there are scientific fundamentalists in the "hard" sciences, who claim that these sciences are all that matters, and decry any attempt to say that there is anything more to life.

Scientism

Scientism—the claim that science is the sole and perfect access to all truth—is a fundamentalist atheist religion, complete with a creed: "Science is the sole route to true, complete, and perfect knowledge" (Peter Atkins, *Galileo's Finger*, page 237) and relic of a saint (the morbid remains of Galileo's finger, depicted in that book immediately after the title page). This approach makes its claims by declamation ("it has to be so") rather than legitimate argumentation, for neither science nor philosophy can establish its main philosophical claims; but it is as dogmatic and closed a belief system as any religion has ever been. It occurs in physics and chemistry, in biology and neuroscience; and proceeds by proscribing what can be legitimately considered the target of enquiry, the methods used, the data allowed, and the kinds of explanation entertained.

Fundamentalism and Humanity

However it often goes much further than simply being anti-religious: it denies the worth of most of what makes civilisation worthwhile. As an example, Professor Atkins believes in "the limitless power of science" (P.W.Atkins in *Nature's Imagination: The Frontiers of Scientific Vision*, ed. J. Cornwell). He writes:

> Scientists, with their implicit trust in reductionism, are privileged to be at the summit of knowledge, and to see further into truth than any of their contemporaries... there is no reason to expect that science cannot deal with any aspect of existence... Science, in contrast to religion, opens up the great questions of being to rational discussion ... reductionist science is omni-competent ... Science has never encountered a barrier that it has not surmounted or that we can at least reasonably suppose it has the power to surmount. ... I do not consider that there is any corner of the real universe or the mental universe that is shielded from [science's] glare (pp. 123, 125, 129, 131).

This is a very clear statement of belief that science can answer questions that are in fact outside its domain of competence. The useful question we can ask is, is Professor Atkins in fact claiming that science can deal with everything of importance to humanity, or rather that anything outside the limited scope of science is unimportant? It appears that the latter is his true position, for he throws out of the window not only theology but also all of philosophy, poetry, and art:

> although poets may aspire to understanding, their talents are more akin to entertaining self-deception. Philosophers too, I am afraid, have contributed to the understanding of the universe little more than poets ... I long for immortality, but I know that my only hope of achieving it is through science and medicine, not through sentiment and its subsets, art and theology (pp. 123, 131).

His frame of reference thus excludes all the highest understandings of the human predicament that have been attained throughout history; he defines reality to be only that which can be comprehended by his narrow view of reductionist science. Indeed he frames his viewpoint so narrowly that it even excludes psychology, all the social sciences, and behavioural biology, for he states "A gross contamination of the reductionist ethic is the concept of purpose. Science has no need of purpose" (p.127). This is the framework within which he claims to consider "the great questions of being". The conclusions he attains are dictated by the self-imposed extraordinarily narrow limits of his analytic scheme.

Professor Atkins is motivated by the power of science to show connections between the disparate, and its ability to show that at its foundations the world is simple (p.126). He rightly wants to see how far this approach can go. However this then becomes a dogma that drives all before it, irrespective of how inadequate it is in some spheres of understanding, and he raises reductionism to a first principle to be adhered to even when it cannot deal with the issues at hand (unlike for example Neil Campbell's superb text *Biology*). Anything that does not fit into this narrow framework is claimed to be self-deception or delusion. His argument against religion, which he characterises as only being based on ignorance and fear (p.124), is essentially that it does not fit into his restrictive scheme.

His whole approach is based on an *a priori* metaphysical position of a fundamentalist nature, which is his ground and starting point. This viewpoint is claimed to be the only metaphysical position compatible with science, which is simply untrue; he can maintain this supposition only by ignoring both the rationally based arguments that carefully consider all the other metaphysical options, and all reasonably sophisticated versions of religious explanation.

One might ask what is the pay-off of this impoverished worldview, which consigns to the dustbin *inter alia* Plato, Aristotle, Kierkegaard, Shakespeare, Dostoyevsky, Tolstoy, Victor Hugo,

Fundamentalism and Humanity

T.S. Eliot? It appears to be two-fold: firstly, one claims absolute certainty (even if this is not attainable)—it is yet another manifestation of the human longing to be free of the metaphysical doubt that in fact faces us. Second, given this view, scientists become the high priests of this barren religion—they are the people with privileged access to omni-competent knowledge. It is their prerogative to judge and dispose of truth in this desolate landscape.

Thus the temptation to scientists to promote this view is the same as has throughout history been the temptation to those claiming absolute knowledge of truth: they can see themselves as superior to their contemporaries. However this no longer works: while a few will follow, the main result is to convince the majority of the public that scientists have little understanding of the real world or of what is valuable. This kind of exaggerated position (whether explicit or implicit) is one of the reasons why anti-science views are presently on the increase in the populace at large.

It is crucial for you to know that this extreme position is not implied by science, and it is not true that all scientists have such a barren and destructive worldview. Scientists don't have to be nerds. Science can be done by people who appreciate the arts and humanities and ethics, and indeed also religion. Some Nobel Prize winners are people of religious faith.

So don't fall for it. Despite all the efforts of scientists, scientific knowledge will always be partial and incomplete, and science cannot comprehend everything of value to humankind. It needs the humility to explicitly recognise that there are limits to what science can offer us, and that other aspects of human endeavour—ethics and aesthetics, philosophy and poetry for example—are also significant; and that religious or spiritual beliefs and experience may also be of great significance. Indeed they are so for most of humanity.

And beware the implicit threat contained in the attitude of scientism, which—as is true of any fundamentalist religion—would like to suppress the thoughts of those who do not agree. I quote from Atkins:

> [Theologians] *have no right* to claim that God is an extreme simplicity ... Maintaining that God is an explanation is an abnegation of the precious power of human reasoning [p.128, my emphasis].

Here you have the clear whiff of the Inquisition: if they have *no right* to think this wrongheaded way, then the strong implication is that they should be stopped from doing so.

Reductionism and Humanity

A further key example is in relation to the nature of humanity. As regards human behaviour and nature, on the one side we have sociologists and anthropologists who say these are totally determined by culture, nothing else matters, and on the other side evolutionary biologists who say they are totally determined by evolutionary history and the resulting genes, nothing else matters. But both these factions cannot be right. In fact both these kinds of causation are significant, and in addition we have the ability to shape ourselves, to a degree, through our own conscious choices. Personal choice matters too. To claim that any of these aspects has no influence is ludicrous, for they clearly all do; but some fundamentalists make this claim. A clear statement of how they have done so historically, and the negative consequences of this stance, is given in the book *The Blank Slate* by Steven Pinker.

What is very interesting is the way that the brain actually develops. The interaction of our brain with the environment, internal and external, is a main factor in shaping the brain. This has to be the case for a simple reason: from the human genome project we now have determined there are about 40,000 genes, roughly, in the entire genome. From these, we have to construct the entire human body. But there are 10^{11} neurons in the brain, each of which has up to 100, maybe even 1000 connections. That is 10^{14} connections. Hence the genome does not contain a fraction of the information necessary to structure the brain, even if each gene is read many times.

What the genome does is set up general principles of structuring the brain and broad functional areas, together with the senses and instinctive and basic emotional functions. All the detailed structuring of the higher brain is governed by the interactions we have with our peers, parents, caregivers, the environment, and with our own minds. So the genetic influence is very important in setting the basic structure, but all the detailed structuring of each of our brains comes through the combination of all these interactions. Consequently genetically identical twins will not have identical brains.

This whole area of genetics and of what makes human identity is becoming more and more important as we get increasing control over the human gene on the one hand and grow in our understanding of the brain on the other. The true ethical crunch is going to come as we increase our ability to interfere with the brain genetically. The kind of problems that face us through our increasing ability to alter how we are, are going to become more serious than ever. If you behave a little differently from everybody else, should you be corrected by altering your brain by use of the latest technology? This brings us *inter alia* to eugenics—the intentional breeding of people to produce specific outcomes—which was started by a Quaker, Francis Galton. Eugenics is one area where science can be used in very dangerous ways.

Neo-Behaviourists and the Mind

The human mind and the question of consciousness is one of the most serious potential points of tension between science and religion; and indeed between science and the fullness of humanity. Let me be very specific about this. There are philosophers, psychologists and neuroscientists who tell us that consciousness is an epiphenomenon: it is not real. We are not really conscious but are machines driven by unconscious computations, so that what we think are conscious choices are not real. To me, this is the one real threat from the scientific side. Let me quote from Merlin Donald's book *A Mind so Rare*:

> Hardliners, led by a vanguard of rather voluble philosophers, believe not merely that consciousness is limited, as experimentalists have been saying for years, but that it plays no significant role in human cognition. They believe that we think, speak, and remember entirely outside its influence. Moreover, the use of the term "consciousness" is viewed as pernicious because (note the theological undertones) it leads us into error ... They support the downgrading of consciousness to the status of an epiphenomenon ... A secondary byproduct of the brain's activity, a superficial manifestation of mental activity that plays no role in cognition (pp. 29, 36)

> Dennett is actually denying the biological reality of the self. Selves, he says, hence self-consciousness, are cultural inventions ... the initiation and execution of mental activity is always outside conscious control ... Consciousness is an illusion and we do not exist in any meaningful sense. But, they apologize at great length, this daunting fact Does Not Matter. Life will go on as always, meaningless algorithm after meaningless algorithm, and we can all return to our lives as if Nothing Has Happened. This is rather like telling you your real parents were not the ones you grew to know and love but Jack the Ripper and Elsa, She-Wolf of the SS. But not to worry ... The practical consequences of this deterministic crusade are terrible indeed. There is no sound biological or ideological basis for selfhood, willpower, freedom, or responsibility. The

notion of the conscious life as a vacuum leaves us with an idea of the self that is arbitrary, relative, and much worse, totally empty because it is not really a conscious self, at least not in any important way (pp. 31, 45).

But this is not in fact what is implied by the science, which has a long way to go before it properly understands the brain, and has made virtually no progress at all in understanding the hard problem of consciousness (however many of the hardliners even deny there is such a problem). I recommend reading *A Mind so Rare* for clarity on these issues. It covers the question of the mind and the reality of our mental efforts and choices very well from a philosophical view. I find Merlin Donald's writings on these topics convincing. And personally, I prefer to run this whole argument the other way round, starting with our daily experience.

Consciousness and conscious decisions are obviously real, because that is the primary experience we have in our lives. This is the basis from which all else—including science—proceeds. It is ridiculous to give up that primary experience on the basis of a fundamentalist theory which ignores this fundamental data. And that theory is not even self-consistent, because if Professor Dennett's mind in fact works that way, then you have no reason whatever to believe his theories—for they are then not the result of rational cogitation by a conscious and critical mind. If that were indeed the case, then the entire scientific enterprise would not make sense.

It is important to reiterate here that despite the enormous amount scientists know about neuroscience and its mechanisms, the neural correlates of consciousness, the different brain areas involved and so on, we have no idea of how to solve the hard problem of consciousness. There is not even a beginning of an approach. So in relation to issues which flow from it, like the question of life after death or reincarnation, I have to be agnostic. But as to the causal efficacy of consciousness, I take that as a given which underlies our ability to carry out science and to entertain philosophical and metaphysical questions. And as a consequence, ethical choices and decisions are real and meaningful.

Existence and Evidence

One of the most common errors in discussing science and philosophy is mixing up existence with knowledge of existence, ontology with epistemology. Things can exist though we cannot conclusively demonstrate that they exist.

There is a very simple example of this from cosmology. Since the universe is 14,000 million years old, the furthest that we can see in crude terms is 14,000 million light years. On most models of the universe, there is material beyond that distance which we cannot see now and indeed that humans never will be able to see. That does not mean these stars and galaxies do not exist; they are simply beyond where we can interact with them, so we cannot conclusively demonstrate their existence by any kind of observation. But there is a very old human tendency, embodied in theories ranging from logical positivism to extreme relativism, to think that things don't exist if we can't *prove* they exist. Not so. This is simply another form of human hubris.

At a deeper level, our inability to scientifically prove the existence of morality or meaning does not prove they are meaningless concepts. The nature of existence goes way beyond the material world. Hints of other levels of existence are there in the existence of physical laws and the discovery of mathematical reality, as well as in the multiple natures of consciousness and the experience of moral dilemmas. But to recognise these aspects and see their implications, you have to be open to the possible existence of deeper layers of reality: to be sensitive to hints of transcendence embedded in everyday physical existence.

Image and Reality

We need to face up to our inability to see reality directly—to only be able to determine its true nature through indirect images and shadows that give hints as to its true nature but are not to be confused with the true thing itself. This has been stated most beautifully by Isaac Penington in 1653 (as quoted in *Quaker Faith and Practice*, §27.22):

> All Truth is shadow except the last, except the utmost; yet every Truth is true in its own kind. It is substance in its own place, though it be but shadow in another place (for it is but a reflection from an intenser substance); and the shadow is a true shadow, as the substance is a true substance.

Facing up to this uncertainty about the natures of existence involves the humility that gives up the claim that we have achieved certainty, and so brings me back to what I was saying about kenosis and self-emptying. I see this as a generic principle with much wider implications than just in ethics.

The Wider Nature of Kenosis

Learning is based on kenosis in the sense that you have to give up your preconceptions about the way things are in order to see things as they really are. Self-emptying is crucial in the artistic endeavours of people who are truly creative as they work on plays, books, sculpture, whatever: they start by shaping the thing in front of them, but at a certain stage it takes on its own personality and integrity. Then the artist's need ultimately is to respond to it, to respect its integrity, and not to impose the self on it. Kenosis is also the basis of community, for that involves giving up one's own needs to some degree on behalf of the welfare of those around.

To take a further very practical example from the Quaker meeting, the question, should you speak in Meeting? If you have an urging to speak, a voice emerging from inner depths, I suggest that the thing to do is to consciously to give up the need to speak, to let go of the need to be heard: and then to listen in the silence, to wait, to hear. Then you can see if the need to speak still remains after you have given it up, and if it does, you can then respond to it.

Discernment

This is the question of discernment. The problem is that throughout history many people have felt strongly that they were being led by God, but you can tell by their actions that in some cases they were being led not by the God of Jesus, but rather by some other God or by their own self-centredness. Many evils have resulted; *inter alia*, the Crusades and the Inquisition were carried out in the name of God, and the policy of apartheid was supported by a group of churches. The crucial issue is discernment, the testing of such urgings to see if they are really the true voice.

There is a clear link with science here, because the strength of science comes through its process of testing to see if its conclusions are true. The real challenge for us is to test our spiritual leadings for their veracity. Quakers have various mechanisms which we have developed precisely for this purpose. One of the reasons why we have to listen to others is the belief that, no matter how irritating they are, they may have something to which we should be listening. They may have seen the Light in a way that we have missed.

The Inner Voice

The idea of spiritual leadings depends on believing that spiritual experience is in some way real. What does one believe about God's action, and the apparent presence of the Inner Light? If one believes Meeting for Worship is more than self-contemplation, there has to be some kind of channel of connection, whatever one wants to call it, where we have some kind of access to the underlying transcendent reality. In my view, one can make a good case for the action of God through impulses and images coming into the human brain, giving a reality to spiritual experience.

There is a fundamental limit in our ability to predict scientifically what will occur at very small scales, because of quantum uncertainty; this central feature of physics allows such intimations to happen in a way consistent with religious belief and without in any way violating current scientific understanding. For a whole host of reasons, in most circumstances that would be the limit of the intervention, otherwise many theological and ethical problems arise. But this does allow for the reality of spiritual life. That is essential if one is to make sense of the spiritual experience of people through the centuries.

The Issue of Evil

As I said earlier, we are partly formed by our environments and experience, including what we learn from others. Our views of God's action and leadings are influenced by our personal experiences. One of the important things here is to what degree we have experienced good or evil. As mentioned before, the problem of evil is a key one—one of the oldest facing religion. This is not the place for any real attempt at a discussion, except to say that if God plans to create conditions allowing people to come into existence who can exercise free will and are able to use that free will to love others and to love God, then that choice constrains what is possible in other ways; in particular you cannot offer independent beings free will and also prevent them from doing evil. Similarly you cannot create physics and biology that will lead to the existence of humanity that can exercise free will, and also have a world without pain and death.

There is great good in the world as well as evil, and one has in the end to believe that the good that will eventually arise will outweigh the pain and suffering that were entailed in getting there. The Christian believes this must be so, because of the image we have been given of God suffering voluntarily on our behalf—freely accepting that suffering in order to create a greater good. That act not only shows the way for us to go, but also shows that God himself follows that way and accepts the suffering. In the end, it is the contemplation of the Cross that is the solution to evil—not in an intellectual sense but in the sense of allowing us to share both the pain and the glory with God. That is the true nature of love.

The Whole Picture

Conflict between science and religion comes when people see things in a partial way, thinking that part of the picture is the whole picture. We need to listen to what both science and religion can tell us in order to understand the whole. Science can help us understand many aspects of reality[2], and in particular see the fine-tuning in physics that allows our existence. That understanding can be very precise, and it can make a huge impression. Our broader experience can give us a relation to spiritual issues with many dimensions. In terms of the beauty of things, I get that by walking in the mountains every Saturday and looking at birds, trees, waterfalls, flowers, clouds, the sea, and all the rest of it. In terms of religious experience, it is what many Quakers have found in the gathered Meeting for Worship.

Awareness of Transcendence

Consequently I like to talk about "intimations of transcendence" —of perceptions of a kind of existence lying behind the surface appearance, which gives a grounding for meaning, morality, and purpose. I think everyday experience gives such hints in many ways. In some sense—not in a scientifically provable way, but as an intuitive kind of feeling—the beauty and glory of what exists is more than is necessary. It did not have to be that way.

In Cape Town one day I was watching the waves coming into Clifton bay, and there were fourteen dolphins surfing there. One of them was doing back somersaults as he surfed. They were simply having fun! There was no purpose in it, it wasn't going to help them survive or get them food. It was just for the joy of it. It was

---

[2]For an integrated view of present day science, see: www.mth.uct.ac.za/~ellis/cos).html

an unnecessary expenditure of energy—and it was joyous and wonderful.

Life is much more than the necessary minimum. The complex web of interactions and values is much richer than any reductionist argument tries to persuade us is the case.

Embracing the Whole

There are many areas where we do not have any answers. We always need to remember that there are limits to what we can know about both science and religion. But both are important to being a fully rounded human being. We need to incorporate both of them. Even if you are not a scientist, it is worth trying to find out about science because it tells us so much. But this does not mean having to deny religion or indeed humanity. The religious life adds an enormously important dimension to humanity, individually and collectively, when approached in a non-fundamentalist way.

Quakers have much to give the world in this regard through their open-minded and non-dogmatic approach that takes seriously both spiritual realities and ethical action that make as a difference in the world. I continue to find our movement an inspirational experience and a welcoming home—a true basis for life lived in the full.

Bibliography

Peter Atkins, *Galileo's Finger: The Ten Great Ideas of Science.* Oxford University Press, 2003. £8.99.

John D. Barrow and Frank J. Tipler, *The Anthropic Cosmological Principle.* Oxford University Press, 1988. £14.99.

Neil A. Campbell and Jane B. Reece, *Biology* (6th edition). San Francisco & London: Benjamin Cummings. £44.99.

John Cornwell (Ed.), *Nature's Imagination: The Frontiers of Scientific Vision,* Oxford University Press, 1995.

Merlin Donald, *A Mind So Rare: The Evolution of Human Consciousness*. New York & London: W.W. Norton, 2001. £10.00.

George Ellis, *Before the Beginning: Cosmology explained*. Boyars/Bowerdean, 1993. Price £9.95.

George Ellis and Nancey Murphy, *On The Moral Nature of the Universe*. Minneapolis: Fortress Press, 1996. £15.99.

Martin Gardner, *Are Universes Thicker Than Blackberries?* New York & London: W. W. Norton, 2004. £9.99.

Jacques Monod, trans. Austryn Wainhouse, *Chance and Necessity: An Essay on the Natural Philosophy of Modern Biology*. Penguin, 1997 (first pubd 1972).

Steven Pinker, *The Blank Slate*. Allen Lane, 2002; Penguin 2003. £8.99

Martin Rees, *Just Six Numbers: The Deep Forces that Shape the Universe*. Weidenfeld and Nicholson, 1999; Phoenix, 2004 £6.99.

Martin Rees, *Our Cosmic Habitat*. Weidenfeld and Nicholson, 2001; Phoenix, 2002. £14.99.

Religious Society of Friends (Quakers) in Britain, *Christian Faith and Practice*. London Yearly Meeting, 1960. Contains the Neave Brayshaw extract; no longer in print. *Quaker Faith and Practice*. Britain Yearly Meeting, 1995; 2nd ed. 1999. £9.50. Contains the Isaac Penington extract.

Richard Weikart, *From Darwin to Hitler: Evolutionary Ethics, Eugenics and Racism in Germany*. New York & Basingstoke: Palgrave Macmillan, 2004. £45.00.